The life and times of Edward the Bacterium

Rachel Haywood

TSL Publications

Acknowledgements

My thanks to Henry and his charming dinosaur tail for giving me
the inspiration for the bacterium and Ezra Williams for allowing me
to use his 'childhood face' as the model for Ezzy.

First published in Great Britain in 2022
By TSL Publications, Rickmansworth

ISBN / 978-1-914245-88-6

This book is written for children to read with their parents. It is about the life of a bacterium called Edward.

It is the second of my 'science books for children'. In the first book about the adventures of a virus called Ezzy, the life cycle of viruses is explained and their impact on humans. This book attempts to help parents explain bacteria to children. Bacteria, like viruses, can be harmful and cause disease and sometimes death. Like viruses they are very small, and not visible to the naked eye. Unlike viruses, they can also have some benefits in the decomposition of waste and protecting the lining of the gut. Some facts about bacteria are provided throughout the book and an appendix at the end gives more detail including some weblinks for further information.

I hope you enjoy reading this book.

Rachel Haywood

Edward was a very unhappy bacterium. He had a lovely tail but no face. His friend Ezzy the virus had a big face but no body. They were always together.

Bacteria and viruses are primitive organisms which can cause disease. Viruses are only living when they infect a 'host' cell to reproduce themselves. Bacteria are bigger than viruses and live in or on a host.

Edward was unhappy because his life was so hard. Ezzy was always having a great time, jumping into people's throats and making them ill.

Viruses need to enter a host cell and take over the cell machinery to make more viruses. New viruses then pass to a new host, and the first host is either killed or kills the virus after being ill.

Ezzy the Virus was like the sun, he was always burning people up. Humans were making it hard for Edward, with their antibiotics. And he was such a useful bacterium.

Some bacteria, like Ezzy who was a virus and not bacterium, could kill humans.

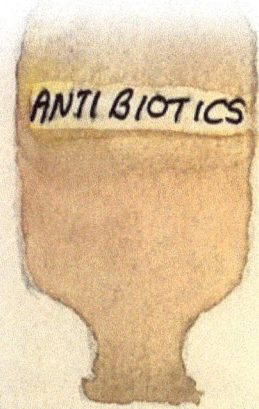

Humans have discovered that bacteria can be killed by chemicals called antibiotics. This stops bacteria casuing infection, sickness and sometimes death. Viruses are not killed by antibiotics.

But Edward was a kind bacterium, who was helpful. He could live in a human's stomach and make it very comfortable. He could also help with turning human waste back into food.

So why was Ezzy the virus successful?

Edward scratched his tail.

Bacteria are not always harmful and can often be useful. They are involved in the decomposition of waste material and 'good' bacteria line the stomach wall and are beneficial preventing 'bad' bacteria from building up and causing disease.

Maybe if he went

on a journey

 he could escape

Ezzy the virus and find a better place

to live. He had a tail and could make

it longer to wriggle. So off Edward

went to explore the world, leaving

Ezzy the Virus

far behind. He laughed because Ezzy was stuck and could not follow him. Edward wriggled off carrying some food in his handkerchief.

Bacteria can often move using a 'tail' and swim to find food but viruses cannot move independently.

He was going to try and
understand how the world
worked. He wanted to make it
a better place.

He came across another floating body, a bit like Ezzy but a lot bigger.

Bacteria can form 'spores' which float in the air and viruses can attach to water droplets which help them transmit to another host. As well as bacteria and viruses, the air is full of spores from plants and the pollen grains which plants use to pass on their genetic material and reproduce.

The pollen grain introduced himself.

He was not very interested

in Edward.

He was happy floating in the wind

and looking for a nice flower to land in and make a new home and family. He did tell Edward a little about his life though. He was born in a plant and then thrown out to find another plant.

Plants make new plants by the 'male' part of the plant producing pollen. This is either carried by the wind or insects to the 'female' parts of another plant and the genetic material combines to produce a seed. Seeds produce new plants in the soil when they find the right conditions.

Plants were giants compared to Edward, the pollen grain and Ezzy the virus. They fed on sunlight, water and carbon dioxide, a

Plants are very different to animals. They cannot move but make their food by combining carbon-dioxide, a gas in the air, with water and they produce oxygen gas as a waste product which animals and humans need to breathe.

gas in the air. They were really clever. They did not need to move or try very hard, like Edward. Edward was not sure he could make a home in a plant. He waved goodbye to the pollen grain and wriggled off again.

Edward could see some cows in the distance. They were happy eating grass.

There were cows everywhere he looked.

He moved a little closer
to a white cow,
which was different
from the black cows.

The cow was enormous.

But Edward soon found a nice wet nose to sit on.

Then the cow licked Edward off its nose and down he went, along a very dark passage and into the cow's stomach! It was so dark and very noisy.

Edward was tossed around on a dark sea. He kept hitting things and was feeling pretty sore. He was sure some of his tail had fallen off.

Poor Edward. He knew he was such a good bacterium but the world did not seem to want him. Where was he now? he wondered. Was he going to get out alive and find a home?

When Edward is swallowed by the cow he travels down the cow's food pipe (oesophagus) into the cow's stomach. Cows actually have four stomachs and bacteria are very helpful to cows in digesting their food. Cows eat grass which is difficult to digest.

Time seemed to pass very slowly and eventually he stopped moving. He seemed to be in a giant cavern full of dark rivers and pink and yellow sandbanks.

He was sat on something

which seemed to be a shelf. All around him he could see the dark rivers, swirling. He was glad to have found his safe place.

He rested after his terrible journey and then fell asleep. When he woke up, everything was still the same. He began to see that he was not alone. On the other side he thought he could see other Edwards!

The stomach is part of the 'alimentary canal' which is a very long tube starting at the mouth and finishing at the bottom. Food in the mouth after chewing pasess along this tube, first in to the stomach and then along the intestines. It is broken down and absorbed through the wall of the intestine through projections called 'villi'. Edward has landed on a villus when he stops moving.

But he could also see some

dark bacteria. They were the 'bad

bacteria'! He knew he did not

belong with the bad bacteria.

He had to get to the other side to where the other Edwards, the good bacteria, were. Then maybe he would be safe and able to make a home. He closed his eyes, held on to his tail, and jumped as hard as he could. He landed with a bang and a thump, squashing

another Edward who was not very

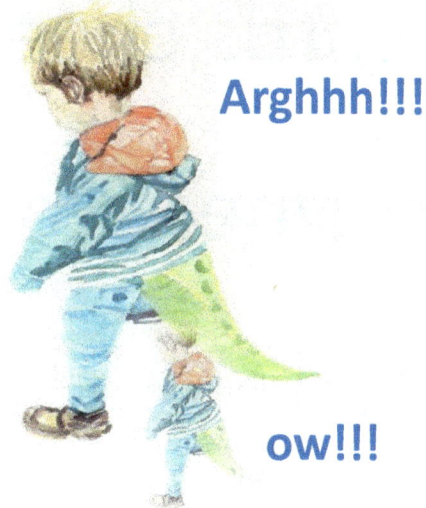

pleased.

Arghhh!!!

ow!!!

There were now more good bacteria

than bad bacteria!

The good bacteria fought the bad bacteria on the other side and won. All the Edwards cheered and they spread everywhere to take the place of the bad bacteria.

Edward, now in a safe place with some friends, lived happily ever after.

Appendix (for parents)[1]

The study of bacteria is a very large subject and information and references can be found online at the time of writing from:

https://en.wikipedia.org/wiki/Bacteria

https://microbiologysociety.org/why-microbiology-matters/what-is-microbiology/bacteria.html

https://www.livescience.com/51641-bacteria.html

In this book I have tried to compare bacteria and viruses at a level children can understand. Bacteria are much bigger, being microns in size, than viruses which are nanometers. Bacteria are just about visible under a light microscope, whearas viruses can only be seen using the more powerful electron microscope. Bacteria are single-celled living organisms that can live independently and are involved in the decomposition of living matter. Viruses cannot live independently and there is debate about whether they are living organisms or not. Inside a host cell they can take over the cellular machinery to produce virus copies. A virus is a strand of DNA inside a protein coat and does not have the cellular structure that bacteria have. Both viruses and bacteria can cause disease in humans, animals and plants but most bacteria are beneficial rather than harmful, whereas viruses usually tend to be harmful.

Humans have discovered antibiotic chemicals [wikipedia.org/wiki/Antibiotic], such as penicillin produced by a fungus [en.wikipedia.org/wiki/Penicillin], which can kill bacteria but they are not effective against viruses. Viruses and bacteria can be deactivated outside the host by ultraviolet light in sunlight, detergents and alcohol in hand sanitisers although bacterial spores may be resistant. Humans and animals, when infected by either pathogenic bacteria or viruses will mount an immune response to attempt to destroy the pathogen. The immune response tends to decline with age and younger people are usually better able to combat bacteria and viruses than older people. The development of vaccines has helped particularly in the fight against viruses:

www.who.int/news-room/feature-stories/detail/how-do-vaccines-work?

A vaccine contains either dead or weakened virus, which gives the immune system a head start in making antibodies to attack the real virus upon exposure. Vaccines against bacteria are rare and we have relied heavily, perhaps too heavily, on antibiotics to kill bacteria. This has resulted in problems of antibiotic resistance as bacteria have developed ways to avoid being destroyed by antibiotics [https://www.who.int/news-room/fact-sheets/detail/antibiotic-resistance].

[1] The sources referenced were checked at the time of publication for scientific accuracy and accessibility of message. No responsibility can be taken for changes or variation in content subsequent to this time.

www.ingramcontent.com/pod-product-compliance
Lightning Source LLC
Chambersburg PA
CBHW062110090426

42741CB00015B/3380

9 781914 245886